Giovani Rezende Barbosa Ferreira
Gustavo F. Morais Dias
Adriano M. Leão de Sousa

Rainwater harvesting

Giovani Rezende Barbosa Ferreira
Gustavo F. Morais Dias
Adriano M. Leão de Sousa

Rainwater harvesting

An alternative for poor communities

ScienciaScripts

Imprint

Any brand names and product names mentioned in this book are subject to trademark, brand or patent protection and are trademarks or registered trademarks of their respective holders. The use of brand names, product names, common names, trade names, product descriptions etc. even without a particular marking in this work is in no way to be construed to mean that such names may be regarded as unrestricted in respect of trademark and brand protection legislation and could thus be used by anyone.

Cover image: www.ingimage.com

This book is a translation from the original published under ISBN 978-613-9-64357-8.

Publisher:
Sciencia Scripts
is a trademark of
Dodo Books Indian Ocean Ltd. and OmniScriptum S.R.L publishing group

120 High Road, East Finchley, London, N2 9ED, United Kingdom
Str. Armeneasca 28/1, office 1, Chisinau MD-2012, Republic of Moldova, Europe
Printed at: see last page
ISBN: 978-620-7-77070-0

SUMMARY

SUMMARY

Currently, the scarcity of drinking water is taking on alarming proportions. Therefore, water must be used and handled rationally. Rainwater harvesting has thus emerged as a water management tool. This study aims to analyse the use of rainwater in the residence of the Cuiarana Experimental Site. As a result, a rainwater harvesting and treatment system was implemented to provide the family with an alternative to the problems of supplying good quality water. The project is located at the Cuiarana experimental site. Specifically at the site's residence, which is approximately 36 m² and is located **in the village of Cuiarana (0º 39' 49.72" S, 47° 17' 03.41" O, 17 m), municipality of** Salinópolis, north-eastern region of the state of Pará. Concern about the cost and implementation of the project was important when drawing up the project, as the study aimed to develop a social technology with a low cost compared to other systems. The cost was R$ 1350.40. The rainwater is collected by the roof of the house and led through gutters that direct it to a duly dimensioned disposal device, which ensures that the first water does not reach the reservoir. After these processes, the water is stored in the cistern and treated. Taking into account the rainfall in the region and the implementation structure, the construction of the project resulted in a rainwater harvesting system with a self-cleaning disposal system and water treatment using an activated carbon and colloidal silver filter. Subsequently, water samples were taken and physico-chemical and bacteriological parameters were analysed using a single sample. The results showed that parameters such as electrical conductivity and ammonia ($NH3^+$) increased after the water was treated. However, pH and temperature decreased. The only parameter that didn't change after treatment was nitrate ($NO3^-$), but the values presented are in line with MS ordinance no. 2.914/11. With regard to total coliforms, the water in the reservoir had an uncountable number, but after treatment this number was sharply reduced, showing 5 CFU/100 mL. The presence of bacteria from this group in the water after treatment is an indication of the inefficiency of the disinfection used or of contamination problems in the distribution network. It is therefore necessary to implement corrective methods. After the water was treated through the filter, turbidity was in line with the drinking water standard. **Escherichia coli** was absent both in the reservoir and after treatment. As this is a pilot project and a single collection, it is very important to monitor the quality of this water over a longer period of time. The search for and implementation of sustainable practices in the Amazon region and their expansion, through social technologies such as the one presented here, aim to provide a better quality of life for communities that lack a drinking water supply.

Keywords: Rainwater. Catchment system. Treatment. Physical-chemical and bacteriological analysis.

CHAPTER 1

INTRODUCTION

97.5 per cent of the planet's water is salty, while fresh water accounts for only 2.5 per cent, of which 68.9 per cent is frozen in the polar ice caps of the Arctic, Antarctica and mountain regions. Underground water makes up 29.9 per cent of the total volume of fresh water on the planet, and only 0.266 per cent of fresh water represents all the water in lakes, rivers and reservoirs (that's 0.007 per cent of the total fresh and salt water on the planet). The rest of the fresh water is found in biomass and in the atmosphere in the form of vapour (TOMAZ, 2003).

The water present in the environment is constantly moving through the hydrological cycle, which refers to the exchange of water between the atmosphere, the soil, surface water, groundwater and plants. Water moves perpetually through stages called evaporation and transpiration, precipitation, runoff and infiltration. This cycle is not always well defined in the various regions of the planet, causing alternating periods of rain and drought, depending on the geography and climatic conditions of the place observed (FIGUEIREDO, 2014).

Water is essential for the life of every plant, animal and human being, and without it the atmosphere, climate, vegetation and agriculture could not be conceived. However, the natural means of transforming drinking water are very limited and slow. That's why water must be used and handled rationally. As a result, rainwater harvesting has emerged as a water management tool, especially for supplying water where this resource is scarce (FERNANDES, 2009).

The scarcity of drinking water is currently taking on alarming proportions. Factors such as the continuous increase in the world's population and the growth of cities and industrial centres are responsible for the increase in demand for quality water. On the other hand, the non-rational and sustainable use of water is causing its scarcity, as it negatively interferes with its hydrological cycle.

There are various forms of water use, but human supply is the one that receives the most attention. The National Water Resources Policy itself, Law No. 9.433/97, obliges priority to be given to water supply over any other (BRASIL, 1997).

The use of treated water, especially in highly populated regions, directly affects existing water sources. The growth of economic activities and the maintenance of the population's quality of life depend on awareness of how to use our water resources rationally. Therefore, investments are needed in technological development and in the search for alternative solutions to increase water supply, as well as actions for efficient demand management, reducing loss and waste rates.

On the world stage, the figures for water use are worrying and the forecast for the future is that world powers will no longer compete for nuclear and oil hegemony, but rather for water reserves to meet the consumption demands of their countries (GOMES, 2011).

The United Nations Organisation presents alarming data (6th WORLD WATER FORUM, 2012) where: 1.7 billion people do not have access to drinking water - equivalent to 18% of the world's population; 2.2 million die each year from water-borne diseases; By 2025, if consumption patterns and high levels of pollution are maintained, two thirds of the planet's population could suffer moderate or severe water shortages; The forecast for 2050 is that only a quarter of humanity will have enough water to meet their basic needs.

Together with public policies to preserve water sources and control pollution, this could satisfy the growing demand for water in large urban centres and also alleviate the problems of scarcity in drier regions and supply inefficiency.

Rainwater harvesting systems are more commonly used in European and Asian countries. These countries offer funding for the construction and use of this system. In Brazil, rainwater has been used in the north-eastern states, due to the great lack of water resources, and this captured water is used as a source of supply (LIMA; MACHADO, 2008).

According to Tomaz (2009), between 1997 and 1999 around 20,000 rainwater storage reservoirs were built in the Northeast. This is an unconventional storage measure to provide water during the dry season. Remember that the water must come from atmospheric precipitation, collected on roofs where there is no circulation of animals, people or vehicles.

The Amazon has a very large quantity of water, concentrating 81 per cent of Brazil's available water resources (ANA, 2010). Around 12 per cent of all surface freshwater in the world. Even so, access to and quality of water are serious problems in the region. In 2008, the North had the highest percentage of municipalities distributing water without any treatment (21.2 per cent). The worst situations are in the states of Pará (40 per cent) and Amazonas (38.7 per cent) (IBGE, 2008). The high availability of water in the Amazon does not reflect its quality.

In the Amazon, rainwater can be an important supply option as a way of making up for the deficit that, ironically, still exists in many places. According to a recent survey by the National Water Agency - ANA (ANA, 2010), around 60 per cent of the municipalities in the state of Pará do not have a wide distribution of treated water.

Faced with the problem of the low availability of drinking water in the Amazon region. A rainwater harvesting system was implemented at the Cuiarana experimental site with the aim of supplying water to local residents affected by the low availability of drinking water.

CHAPTER 2

OBJECTIVE

2.1 General Objective

The aim of this study is to analyse the use of rainwater at the residence of the Cuiarana Experimental Site by implementing a rainwater harvesting system to provide an alternative source of drinking water.

2.2 Specific Objective

a) Carry out a hydrological study of the region where the catchment system is located;

b) **Implement / build a pilot rainwater harvesting system;**

c) Evaluate the use of rainwater;

d) Analysing the quality of the rainwater collected;

e) Present the feasibility of implementing and expanding the project.

CHAPTER 3

THEORETICAL FRAMEWORK

3.1 Water Availability in Brazil

Brazil is the fifth largest country in the world with 8.5 million km^2 divided into five geographical regions, totalling 5565 municipalities and twelve hydrographic regions. It receives an abundant amount of rainfall, which varies over more than 90 per cent of its territory (ANA, 2010). Brazil is one of the richest countries in terms of water resources on the planet: 12 per cent of the planet's fresh surface water is in Brazilian territory, while in regions of Europe, such as Portugal and Spain, as well as the Middle East and much of Africa, water scarcity is chronic (CAMPOS; AZEVEDO, 2013).

According to Matos (2007), Brazil is rich in water availability, but there is temporal and spatial variation in flows. River basins located in areas with low water availability and high water resource use face scarcity situations and a series of social problems caused by the lack of water.

The country's vast territory allows for different climatological and hydrological regimes, which can be exemplified by the abundance of the Amazon River, the largest river discharge in the world, while the semi-arid northeast has serious problems with droughts and dry spells.

There is a great diversity of situations in Brazil. The north and centre-west regions have an abundance of water, with 89% of the country's surface water potential, but only 14.5% of Brazilians live in these regions, which have a water demand of 9.2% of the national total. Meanwhile, the remaining 11 per cent of water potential is spread across the northeast, south and southeast, where 85.5 per cent of the population and 90.8 per cent of the country's water demand is located (IBAMA, 2002).

In addition to surface water, it is also worth mentioning groundwater, which also has huge volumes and great potential for use in Brazil. There are 112,000 km of permanent groundwater reserves[3] (IBAMA, 2002) and UNESCO studies estimate that there are around 10 per cent of the 250 million wells in operation in the world

(REBOUÇAS, 1999).

According to Hespanhol (2003), water has become a limiting factor for urban, industrial and agricultural development, causing significant changes in people's quality of life. The use of rainwater has therefore become a viable way of minimising urban problems and the threat of possible social conflicts.

3.2 Rathering Rainwater

Harnessing rainwater is a very old practice. It began in the Middle East, dating back to 850 BC, when King Mesha of the Moabites left an engraving on the Moabite Stone of his desire to install a reservoir for the utilisation of rainwater in each house. It has spread and gained momentum over the centuries in countries such as the United States, Japan, Germany and Australia. These countries offer funding for the construction of rainwater harvesting areas. The United States, for example, has more than 200,000 rainwater harvesting tanks. In Japan, in the city of Sumida, rainwater is an alternative to ensure greater security of supply in potential emergency situations. Currently, these countries use a dual cold water distribution system that generates water savings of more than 30 per cent: one for potable purposes and the other for non-potable purposes. The latter are mainly used for sanitary basins (TOMAZ, 2003).

The use of rainwater is simple and easy to understand. At its base, it is an alternative water supply model that makes use of impermeable areas, including roofs, slabs, pavements, among others, which aim to collect the product of rainfall in their own reservoir(s) and distribute it from there (VELOSO, 2012).

Rainwater is an excellent alternative for supplying drinking water during periods of drought, as it has an appropriate level of small-scale technology, affordable costs, immediate results and is simple to maintain.

The feasibility of implementing a rainwater utilisation system depends essentially on the following factors: rainfall, catchment area and water demand. In addition, the design of such a system must take into account local environmental conditions, climate, economic factors, the purpose and uses of the water, so as not to standardise technical solutions.

In order to structure a rainwater harvesting system, it is essential to have some components such as: a catchment area, gutters and conductors, a device for disposing of the first water, sieves, a reservoir and, depending on the use of the water collected, treatment.

The catchment area is made up of roofs that can be made of ceramic tiles, where the roof can be sloping, slightly sloping or flat. Roofs can be flat, moderately sloping or intensely sloping. Flat roofs require internal drains or gutters along the perimeter, while moderately sloping roofs drain the water off the roof.
more easily and intensely sloped roofs produce faster runoff. A roof with a slight slope increases the catchment area.

Gutters and collectors can be made of PVC or metal and sized according to the rainfall in each region. These components are used to separate the initial rainwater, which contains excessive concentrations of organic matter and dissolved solids deposited by the wind, birds and insects (TOMAZ apud TORDO, 2004).

The disposal device must be located before the reservoir and must be sized according to the catchment area. This process is necessary to guarantee the safety of users, since the first water contains a series of microorganisms, some natural, carried by the wind, and others that have proliferated in the environment and can be harmful to human health, such as some pathogens (TORDO, 2004).

Another essential component for efficient rainwater harvesting is the 0.2mm to 1.0mm mesh screens used to prevent suspended solids from entering the water tank.

The reservoir must be sized to meet the required demand, which is defined by the designer, and the volume of the reservoir will depend on this demand, the catchment area and the rainfall characteristics in the region, which requires studies of the historical and synthetic rainfall series in the region where the rainwater utilisation project will be carried out.

Depending on the use of this water, it needs to be treated, taking into account chemical, physical and biological agents in order to guarantee the quality of this water resource and the health and well-being of the end consumer.

Some precautions are necessary when selecting the location of this reservoir and the catchment area, such as: keeping away from possible contaminants (septic tanks), areas with birds and vegetation, cleaning these areas by removing foliage from the roof and gutters, washing the reservoir, and in relation to the treatment system, the useful life of filters and chemicals should be observed.

3.3 Rainwater Harvesting in Brazil and the World

Harnessing rainwater is already a widespread practice in many parts of the world. In some regions, its use is extremely necessary and it is an important way of accessing water. In other cases, it is used as a way of preserving water sources and/or saving on the consumption of treated water.

Nowadays, the use of rainwater resources is widespread in developed countries, even with strong legislation on the issue. Japan, the USA, Germany and Australia are examples of countries that use rainwater in a variety of applications: from drinking it to meet drinking needs to less noble purposes such as washing and watering gardens (VELOSO, 2012).

In Brazil, the first report of rainwater utilisation is probably a system built on Fernando de Noronha Island by the US army in 1943 (GHANAYEM, 2001 apud PETERS, 2006).

There are currently few reports of rainwater harvesting for reuse in Brazil, given the relatively large availability of other sources of supply. Rainwater harvesting has been practised on a larger scale mainly in the Northeast, due to the problem of water scarcity that characterises part of the region. In July 2003, the Training and Social Mobilisation Programme for Coexistence with the Semi-Arid: One Million Rural Cisterns - P1MC - began, with the aim of benefiting around 5 million people in the semi-arid region with drinking water through the construction of cisterns.

According to Tassi (2014), Brazilian cities that have gone through the process of implementing an Urban Drainage Master Plan (PDDrU) have been demanding that new developments build a system to control the increase in flows generated by waterproofing. The PDDrU themselves suggest that, among other structures, reservoirs be installed for the temporary storage of rainwater, as is the case in the cities of São Paulo, Curitiba, Santo André, Porto Alegre, Rio de Janeiro, Caxias do Sul, Salvador, Guarulhos, among others.

In Brasilia/DF, District Law No. 4.181/08, which creates the Rainwater Harvesting Programme, states that the Executive Branch of the Federal District will encourage and support the installation of water tanks or reservoirs, with partially removable lids, to collect and store atmospheric precipitation in houses and buildings, both public and private, with more than 200 m² of built area. The granting of habitation permits for buildings begun after this law came into force is subject to proof of compliance with the provisions of this law.

In São Paulo/SP, Municipal Law No. 12.526/07 makes it compulsory to build a reservoir to store rainwater collected on roofs and pavements on plots, built or not, that have a waterproofed area of more than 500 m² .

In Rio de Janeiro/RJ, Municipal Decree No. 23.940/04 makes it compulsory to build rainwater storage tanks in developments with a waterproofed area of more than 500 m² .

In Guarulhos/SP, Municipal Law 6.511/09 instituted the Municipal Programme for the Rational Use of Drinking Water, one of the objectives of which is to adopt measures to regulate, oblige and supervise the installation of reservoirs to capture rainwater and/or drainage water in new buildings in the city. The civil construction project for new buildings with a roof area equal to or greater than 250 metres² must present technical solutions to be applied to the buildings in order to include the installation of reservoirs for collecting rainwater and/or drainage.

In Cascavel/PR, Law No. 4.631/07 of the municipality of Cascavel in the state of Paraná instituted the Municipal Programme for the Conservation and Rational Use of Water and Reuse in Buildings. The aim of this programme is to institute measures that induce conservation, rational use and the use of alternative sources for water collection and reuse in new buildings, as well as raising awareness among users of the importance of water conservation. As an incentive and a discount on the Urban Land

and Property Tax (IPTU) for properties that fall within the scope of this law, in relation to the industrial sector, these must adopt the reuse of rainwater for their activities.

In Curitiba/PR, Municipal Law No. 10.785/03, which instituted the Programme for the Conservation and Rational Use of Water in Buildings (PURAE), aims to institute measures that induce conservation, rational use and the use of alternative sources for water collection in new buildings, as well as raising awareness among users about the importance of water conservation. The municipal legislation suggests that all new buildings install reservoirs with a minimum volume of 500 litres.

In Porto Alegre/RS, Municipal Law No. 10.506/2008 - The Programme for the Conservation, Rational Use and Reuse of Water aims to promote the necessary measures for the conservation, reduction of waste and use of alternative sources for the collection and use of water in buildings, as well as raising awareness among users of its importance to life. Mandatory for commercial and industrial buildings with a roof area of more than 500 metres^2 .

3.4The   Use of Rainwater in the Amazon Region

The utilisation of rainwater in the Amazon is in its infancy. Scientific literature is still scarce. There are only a few isolated experiments using rainwater for drinking and non-potable purposes, supported by institutional or business projects (VELOSO, 2012).

In the state of Amazonas, one state initiative that deserves to be highlighted is the Programme for Home Sanitation Improvements, Rainwater Harnessing and Storage - Prochuva. Initially developed by the Amazonas State Secretariat for the Environment and Sustainable Development (SDS) and later in partnership with the National Health Foundation (FUNASA), the programme has been running since 2006 and consists of the distribution of an infrastructure kit containing the essential parts of the system: gutters, pipes and a water reservoir. The aim is, through the use of rainwater resources, to benefit communities affected by river droughts that have no domestic water supply system, as shown in Figure 1 (SDS, 2007).

Figure 1 - House benefiting from Prochuva.
Source: AMAZONAS (2007).

Prochuva prioritises populations living in Conservation **Units** in the state of Amazonas. Communities located in the channels of the Purus, Solimões, Amazonas and Madeira rivers, comprising a total of 15 municipalities, including: Maués, Parintins, Nhamundá, Borba, Novo Aripuanã, Manicoré, Beruri, Anori, Codajás, Coari, Tefé, Manaus, Iranduba and Manaquiri, are the objects of this action, which aims to benefit 9,413 residents (SDS, 2007).

By 2007, 1,839 domestic systems had been set up, as well as 108 community systems. The community being supported is spread across different regions, including: the Piagaçu-Purus sustainable development reserves, the Madeira River and the Uatumã River, the Nhamundá and Maués state forests, localities in the Lower Amazon and Lower Solimões sub-basins and municipalities hit by the great drought of 2005 (AMAZONAS, 2007).

Also in Amazonas, the Centre for Analysis, Research and Technological Innovation Foundation (FUCAPI) developed the "Clean water for small communities in the Amazon" project in 2010. The intention was to provide an alternative for communities in the interior of the state that suffer from problems supplying good quality water during the period when rivers are flowing, by collecting and treating rainwater (VELOSO, 2012).

The prototype developed (Figure 2) does not require electricity. It uses two gutters to collect rainwater. As it flows down the gutters, the water passes through a first filter to remove debris such as stones and leaves; there is a valve to dispose of this first, dirtier water. The water then goes through a pebble filter, into a first tank and then through a sand filter, an activated charcoal filter and a screen filter, before being stored in a 500-litre tank, where it can be used.

Figure 2 - Prototype developed by FUCAPI.
Source: FUCAPI (2012).

For Veloso (2012), it is clear that this is a much more careful initiative in terms of system efficiency, since it is concerned with the quality of the water to be supplied. This is a major step forward for the Amazonian scenario, since it recognises that it is not enough to have water, it needs to be drinkable. It is true that it has not yet reached potability in its treatment, but the path towards this is clear.

In the state of Pará, the first recorded experience of harnessing **rainwater took place in 2004, with the implementation of the "Clean water is life" project. "The** system featured a cistern along the lines of those built in the **northeastern** semi-arid region" **(ROSA, 2011), see Figure** 3. It was a multi-institutional initiative between the Bible Society of Brazil (SBB), the Ministry of Agrarian Development (MDA), the Dom Helder project and Diaconia.

Figure 3 - Cistern on Ilha Grande.
Source: VELOSO (2012).

The cistern, with a capacity for 16,000 litres of water, was built on Ilha Grande, located in the southern part of Belém. The system was set up to serve the community in general and the São José Elementary School and follows the model of the cisterns of the One Million Cistern Programme - P1MC, which was set up in the Brazilian semi-arid region and is currently inoperative (VELOSO, 2012).

The capital of Pará also has other experiences, mainly on islands. In **2006, the "Água em Casa, Limpa e Saudável" (Water at Home, Clean and Healthy) project was set up in the island region,** promoted by Cáritas Metropolitana de Belém - CAMEBE, a private, non-profit religious association that seeks to promote charity in a broad way and to integrate and strengthen human dignity (VELOSO, 2012). According to CAMEBE (2007) **"the** main objective of the initiative is to implement rainwater collection and treatment systems **for families on the islands of Belém who live with a lack of drinking water availability."** The project consists of storing rainwater, with no initial disposal, which is directed to the reservoirs through gutters and pipes installed in the houses, and then used by the families, as shown in Figure 4.

Figure 4 - Model of CAMEBE's rainwater harvesting system.
Source: SOUZA (2012).

CAMEBE has been seeking partnerships to develop and improve its activities. In this way, with the aim of studying the health impacts of using rainwater, IFPA, between 2008 and 2011, and the University of Amazonia - UNAMA, in 2010 and 2011, provided scientific support through axes of research focused on the development of technologies, the evaluation of impacts on the health of residents, epidemiological aspects, anthropological studies and economic effects on the population (SOUZA, 2012).

The Federal University of Pará - UFPA, through the **"Harnessing Rainwater in the Amazon"** Research Group **of the Environment Centre** (Postgraduate **Programme** in Natural Resource Management and Local Development in the Amazon - PPGEDAM/NUMA) and the Institute of Technology (Civil Engineering Programme/ITEC), **since 2008, has been conducting a study that aims to "develop models of supply systems** and construction projects for social housing using rainwater as an alternative in order to enable **Amazonian riverside communities to have** access to **drinking water"** (OLIVEIRA, 2009 apud VELOSO, 2012).

Research on the Grande and Murutucu islands has been funded by the Pará State Research Support Foundation (FAPESPA) and the National Council for Scientific and Technological Development (CNPq), and will soon expand to the Combú and Maracujá islands (Figure 5). Together, these islands cover an area of 39.6

km² , all in the Guamá River.

Figure 5 - Projects of the Rainwater Harnessing in the Amazon Research Group.
Source: (a) VELOSO (2012); (b) GONÇALVES (2012).

The Federal Rural University of Amazonia's **project "Promoting** Sociobiodiversity: Environmental Restoration with Income Generation in **Riverine** Communities in **Amazonia" won the 2014 ANA Prize in the "teaching" category, and the** **"Melipoliniculture Project", winner of the** 2011 Santander Solidarity University **Prize,** coordinated by Professor Vânia Neu (ISARH/UFRA), stood out for promoting significant changes in the lives of the riverine community of Ilha das Onças, in the municipality of Barcarena (UFRA Notícias Hoje, 2014).

The project has already brought about significant changes in the community's daily life. Among the main ones are the implementation of ecological toilets; paper recycling; an increase in income through environmental restoration (and consequent increase in the productivity of native plant species, especially açaí), the introduction of meliponiculture, with honey production and cisterns that capture rainwater, as can be seen in figure 6 (UFRA Notícias Hoje, 2014).

17

Figure 6 - Cistern used to collect rainwater for the project.
Source: UFRA News Today (2014).

Thus, initiatives to use rainwater for water supply come from a wide range of social actors: public bodies, non-governmental organisations, residents' associations, teaching, research and extension institutions and even individual river dwellers.

The context reveals that utilising rainwater on Amazonian islands as an alternative water supply system is an intelligent local management strategy. Due to their geographical peculiarities, most of these regions do not have their own water supply system for their residents.

Thus, with all this history in the Amazon region of implementing rainwater harvesting projects, it aroused interest in looking for and inserting yet another alternative in these systems within the region.

CHAPTER 4

CASE STUDY

4.1 Study Area

The work was carried out at the Cuiarana experimental site (**0°39'49.72"S, 47°17'03.41"O, 17 m**), belonging to the Federal Rural University of Amazonia (UFRA), in the municipality of Salinópolis, in the north-east of the state of Pará (Figure 7), with an area of 25.8 hectares. The site is home to a house measuring approximately 72 metres² (Figure 8). Access to the area is via the BR-316 and PA-124 motorways, 220 km from Belém, the capital of the state of Pará.

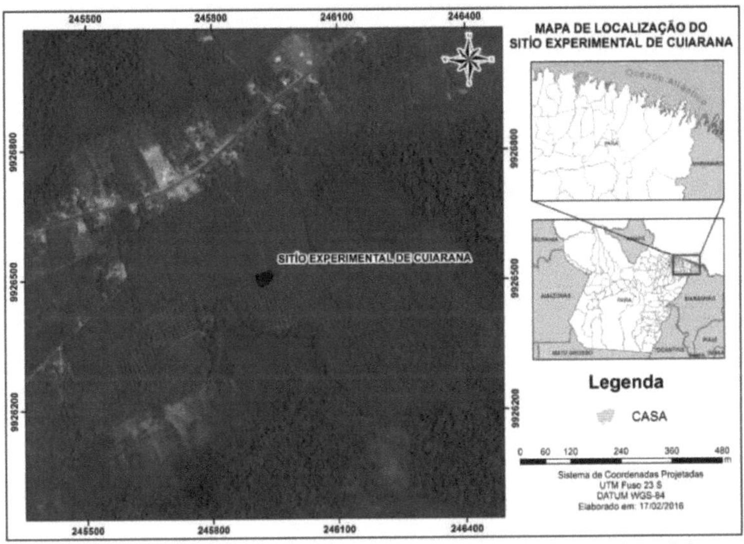

Figure 7 - Geographical location of the Cuiarana - PA experimental site.
Source: Author (2016).

Figure 8 - House at the Cuiarana Experimental Site.
Source: Author (2016).

4.2 Rainfall data

According to the Koppen classification, the predominant climate type in the region is Aw, with a low temperature range and average annual rainfall of 2,100 mm, 90% of which is distributed between the months of January and June (RODRIGUES, 2012).

Using data from the Micrometeorological station at the Cuiarana Experimental Site, we can see the accumulated rainfall in the locality in recent years (Table 1) and its distribution over the months by means of the monthly average in the years observed (Figure 9).

Table 1 - Accumulated rainfall from 2011 to 2014.

Year	2011	2012	2013	2014
Precipitation (mm)	3017,28	1550,16	1856,48	2831,32

Source: Research data (2015).

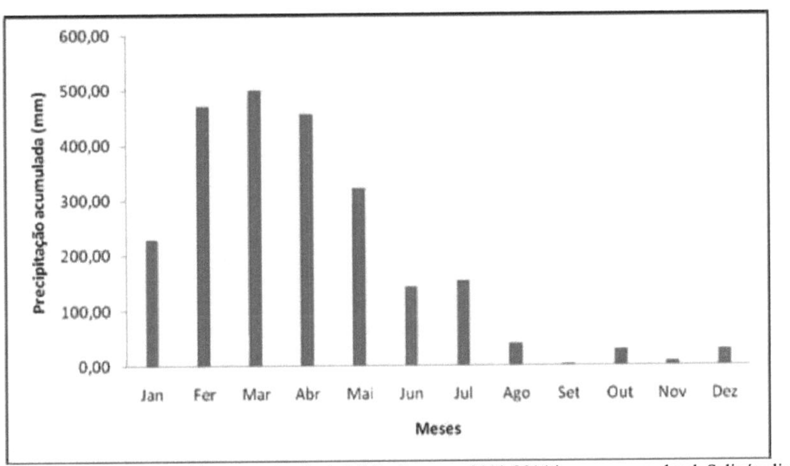

Figure 9 - Average monthly accumulated rainfall for the years 2011-2014 in a mango orchard, Salinópolis, PA.
Source: Author (2015).

From the rainfall data, it is possible to observe the behaviour of precipitation in the locality, so the region has satisfactory levels, according to Table 1 it reached values of 1550.16 mm to 3017.28 mm in the years 2011 to 2014. Figure 9 shows the seasonality of the Cuiarana region, where January to July is considered the wettest period and August to December the least rainy. It is of great importance to know the behaviour of rainfall in the region in order to base the project to be implemented.

4.3 Materials Used to Install the Cistern

The materials used to build the rainwater harvesting project and the total cost of implementing the harvesting system can be seen in Table 2. Concern about the cost and implementation of the project was of great importance in the design of the project, as the study aimed to implement a social technology, which in this case is a simple rainwater harvesting system, without the need for pumps and hydraulics and the use of electricity, thus aiming for a low cost, which was R$ 1350.40, compared to other systems such as Figueiredo (2014), Lima and Machado (2008) and Teston (2012), which cost R$ 7.117.00, R$8,224.00 and R$2,997.21 respectively, and its applicability in different scenarios for different regions.

Table 2 - List of materials and cost to build the project

Item	Quantity	Unit	Materials	Unit Price (R$)	Total Price (R$)
1	1	L	**500 litre polyethylene water tank**	200,00	200,00
2	30	m	Wooden slats and plywood	50,00	250,00
3	2	Kg	Nails	7,50	15,00
4	12	m	PVC pipe 150 mm	160,00	320,00
5	2	Unit	PVC cap 150 mm	34,00	68,00
6	5	m	0.5 mm protective screen	3,00	15,00
7	18	m	PVC pipe 75 mm	66,00	198,00
8	1	Unit	PVC cap 75 mm	16,00	16,00
9	11	Unit	PVC TE 75 mm	8,90	98,00
10	4	Unit	75 mm 90° knee	5,50	22,00
11	12	m	PVC pipe 25 mm	13,00	26,00
12	1	Unit	PVC TE 25 mm	Rye	1,10
13	4	Unit	PVC 90° knee 25 mm	1,05	4,20
14	1	Unit	PVC ball valve 25 mm	11,20	11,20
15	3	Unit	Tube adhesive (45g)	8,33	25,00
16	1	Unit	Water tank with filter (6 litres)	69,00	69,00
17	1	Unit	Activated carbon and colloidal silver filter	11,90	11,90
			Total		**1350,40**

Source: Author (2015).

For the physical-chemical analysis, the Hanna Multiparameter Probe and Phmeter were used to provide data to be analysed on the quality of the rainwater captured and treated by the system.

4.4The   **Implementation of the Rainwater Harvesting and Treatment System**

A project was drawn up in which the water is collected by the roof of the house and transported by gutters that direct it to the disposal device, causing the first water to be discarded. After these processes, the water is stored in the cistern, which is elevated, thus avoiding the expense of hydraulic pumps to suck the water contained in the reservoir into the treatment system, where the system's cycle comes to an end. The entire process, from rainwater collection to treatment, is carried out using gravity and hydraulic tests.

4.4. 1Catchment Area

The residence has a catchment area of 72 m^2 , but the area used to capture water was 36 m^2 , which is covered by ceramic tiles. It was noted that there is a lot of vegetation around the house, making it necessary to take care to clean the roof.

4.4.2Water transport   and distribution

To transport and distribute the captured rainwater, a system of 150 mm PVC gutters was produced, along with 0.5 mm mesh sieves used to retain suspended solids present (Figure 10). The main aim was to retain leaves that could enter the system.

Figure 10 - Chute and sieve system
Source: Author (2015).

Maintenance/cleaning of the system's gutters and pipes will directly influence its efficiency. This maintenance will prevent blockages and reduce the risk of contamination.

The water transported by the gutters is directed through 75 mm PVC pipes to the disposal device, where the function of this component is to separate and dispose of the first heavy rainwater that washes away the atmosphere, the roof, gutters and pipes.

4.4. 3Disposal system

In the project, the DESVIUFPE disposal device was implemented. This device was created by civil engineer Júlio César Azevedo and his master's supervisor Sávia Gavazza on the UFPE campus in Caruaru, a town in the Agreste region of Pernambuco. The device consists of a set of PVC pipes and fittings that receive the first litres of rainwater compromised by dirt on the roofs of homes. The construction of the DESVIUFPE is simple: based on the rainwater catchment area, a calculation is made for the construction of the disposal device. Thus, the disposal system is sized in the proportion of 1 m^2 for the disposal of 1 L of water (Figure 11).

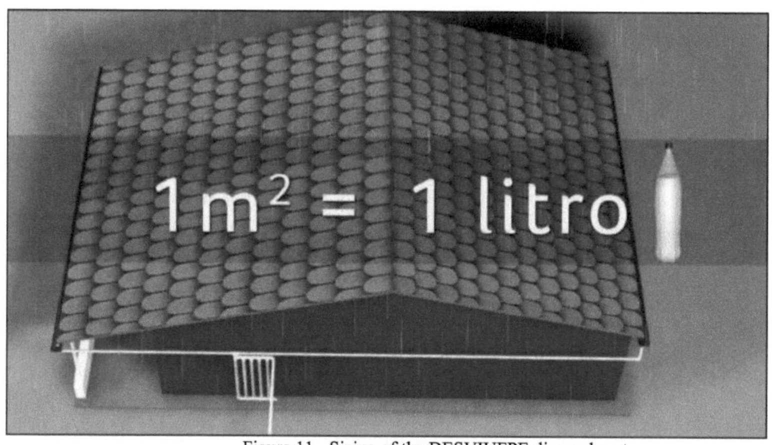

Figure 11 - Sizing of the DESVIUFPE disposal system.
Source: UFPE (2015).

Thus, for every 1 m^2 of catchment area, it is necessary to discard 1 L of rainwater, so for our study, it was necessary to discard 36 L of water.

DESVIUFPE is built using PVC pipes and fittings. To build the system, it is necessary to calculate the number of pipes and fittings that will be needed to supply the volume of water to be discharged, and this will depend on the dimensions of the pipe to be worked on. In the case of this study, 75 mm PVC pipes were used. As it is a cylinder, using spatial geometry, it is possible to calculate the volume of the cylinder and the water stored using Equation 1.

$$V = \pi r^2 h$$

Eq. (1)

Where:

V - Cylinder volume (L)

π - Pi, referring to the ratio between the perimeter of a circle and its diameter;

r - Radius, equal to half the circumference;

h - Height (m)

Therefore, in the project, the calculated volume will first be given in the unit

24

of cm^3 , and then converted into Litres (L). As variables we have π equal to 3.14; the radius (cm) is half the size of the tube equal to 3.75 cm (35 mm); and the height (cm) equal to 150 cm (1.5m), as calculated below:

$$V= 3.14 \times (3.75)^2 \times 150$$

$$V \approx 6623 \text{ cm}^3$$

As 1 litre equals 1000 cm^3 , we have that:

$$1 \text{ L} \text{ ------------------- } 1000 \text{ cm}^3$$

$$V(L) \text{ ----------------- } 6623 \text{ cm}^3$$

$$V(L) = 6.62 \text{ Litres/tube}$$

It was then observed that each 75 mm PVC pipe, 150 cm high, has a volume of approximately 6.62 litres. With this data, it was possible to calculate how many pipes were needed to build the DESVIUFPE system for the disposal of 36 litres of water, as shown in the calculation below.

Like this,

QuantityVolume

$$1 \text{ Tube } 6 \text{ -------------------- } .62 \text{ L}$$

$$X \text{ --------------------------- } 36 \text{ litre tubes}$$

$$X = 5.44 \text{ tubes}$$

or

$$X \approx 6 \text{ Tubes}$$

With this quantity, we were also able to size the fittings used in the system, in this case 75 mm 90° elbows and 75 mm TEs, as they vary according to the number of pipes. In the study, **4 75 mm 90° elbows and 11 75 mm TEs** were used to build the **DESVIUFPE.**

As a final product, we obtained the DESVIUFPE disposal structure, which was used in the diversion system coupled to the rainwater collection system at the residence of the Cuiarana Experimental Site, Salinópolis-Pa (Figure 12).

Figure 12 - DESVIUFPE disposal structure at the residence of the Cuiarana Experimental Site, Salinópolis-PA.
Source: Author (2015).

After each rainfall event, it must be emptied through a discharge pipe, which must be closed again to allow the automatic diversion of the first waters of the next event to operate.

In the study, the water captured in the disposal system was used for non-potable purposes such as washing pavements, tools, irrigation and so on.

1.1.4 Storage Tank

The cistern can be supported on the ground or buried, and whenever possible it should be located close to the points of consumption to reduce the distance the water has to be transported.

The most commonly used materials are concrete, masonry, ferro-cement, galvanised metal, fibreglass and polypropylene (HAGEMANN, 2009).

The project used a polyethylene cistern with a capacity of 500 litres, which has smooth internal surfaces that make it easier to clean and does not require screws and ties during installation, thus guaranteeing even better sealing and water conservation. The dimensions of the tank can be seen in Figure 13 and Table 3.

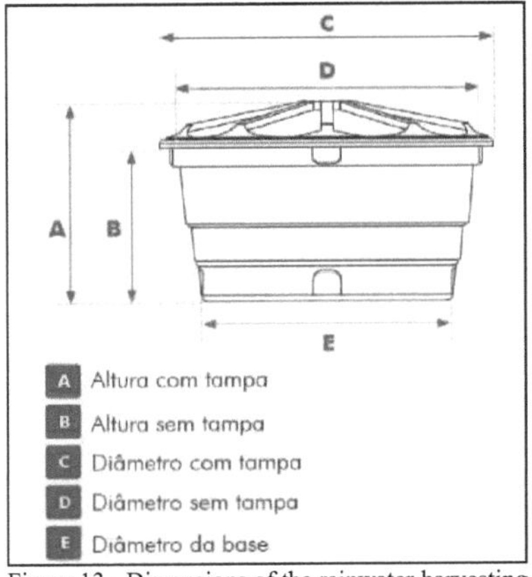

Figure 13 - Dimensions of the rainwater harvesting system reservoir.
Source: FORTLEV (2015).

Table 3 - Polythene tank dimensions

Capacity (L)	Dimensions (m)				
	A	B	C	D	E
500	0,72	0,58	1,24	1,22	0,95

Source: FORTLEV (2015).

Inside the reservoir, the distribution of the water collected through the 75 mm PVC pipe has a turbulence reducer, which was built with the aim of preventing turbulence in the cistern, as it could cause all the dirt settled at the bottom to be turned over, thus impacting on the water quality, as can be seen in Figure 14.

Figure 14 - Turbulence Reducer.
Source: Author (2015).

After the reservoir, the distribution of the collected water is directed to the home, where it has two paths, one that connects to a tap that will serve for non-potable uses with the aim of not wasting the water resource, and the other to the treatment system.

1.1.5 Water Treatment

After all these steps, the water collected by the system is directed into the home, where it is placed in a reservoir with an activated carbon and colloidal silver filter (Figure 15), which holds approximately 6 litres, so that the water can be treated.

Figure 15 - Rainwater treatment structure.

Source: Author (2015).

The filter has a wall made of microporous ceramic material that filters the water by gravity, without the addition of chemicals, reducing impurities and retaining solid particles, which guarantees drinking water free of micro-organisms (STÉFANI, 2015), as can be seen in Figure 16.

Figure 16 - Activated carbon filter with colloidal silver.

Source: Stéfani (2015).

The filter candle has a triple action, where it is dechlorinating and sterilising, with a coating of colloidal silver, applied to the inside of the candle, together with activated carbon, working to filter and reduce odours, flavours and chlorine content, as well as reducing the presence of bacteria in the water. The silver layer applied internally penetrates through the candle's pores, preserving its sterilising properties (STÉFANI, 2015).

4.5 Collection and Preservation of Samples for Water Analysis

Water samples were collected and preserved in accordance with the National Guide for Collecting and Preserving Samples of Water, Sediment, Aquatic Communities and Liquid Effluents, an initiative of CETESB (Companhia Ambiental do Estado de São Paulo).

During the collection stage, one of the fundamental procedures for guaranteeing that the sample's characteristics are maintained is its correct packaging. Therefore, the collection bottles must be resistant, chemically inert, well sealed and preferably easy to handle both for collection and cleaning. The most commonly used bottles are borosilicate glass, amber borosilicate glass (used to prevent photodegradation of the sample) and polyethylene. The bottles must be clean and always sealed; to clean the bottles, detergents should be used to wash laboratory glassware, soak them for 24 hours in a 10% nitric acid solution (or hydrochloric acid for nitrogen ion analyses) and finally rinse them with deionised or reverse osmosis water (COMPANHIA AMBIENTAL DO ESTADO DE SÃO PAULO, 2011).

The water temperature, pH, conductivity, salinity and dissolved solids content, whose characteristics are not maintained with preservation, were determined in situ. The other parameters should be analysed as soon as possible in the laboratory in order to minimise volatilisation or biodegradation between sampling and analysis.

4.6 Parameters Analysed

Physico-chemical and bacteriological parameters were analysed. The physical-chemical parameters collected were chosen in order to visualise their relationship with the bacteriological parameters.

According to Von Sperling (2005), the presence of suspended solids can provide protection against pathogens. Suspended solids are the constituent responsible for turbidity, which represents the degree of interference with the passage of light through water. Particulate matter can be used as an indication of the likely presence of microorganisms. Based on this observation, it was necessary to analyse the turbidity and suspended solids in the samples.

The parameters analysed are shown in Table 4 and were assessed according to the methodology recommended by the Standard Methods for the Examination of Water and Wastewater (APHA, 2005). The samples collected were sent to the Hydrochemistry Laboratory at the Geosciences Institute (IG) of the Federal University of Pará (UFPA), where the water quality analyses were carried out.

Table 4 - Parameters analysed and methodology used.

PARAMETERS	METHOD
Electrical Conductivity	ORION portable conductivity meter
Salinity	Hanna Probe
PH	ORION portable phantom
Water Temperature	Temperature sensor attached to the phantom
Turbidity	Portable turbidimeter from HACH
Ammonia	Nessler Spectophotometric Method
Nitrate	Cadmium reduction method - Spectrophotometric
Total coliforms	NKS - Filter Membrane Method
Escherichia coli (E. coli)	NKS - Filter Membrane Method

Source: Author (2015).

CHAPTER 5

RESULTS AND DISCUSSIONS

At the end of the study into rainwater harvesting and utilisation at the residence on the Cuiarana Experimental Site in Salinópolis-PA, taking into account the rainfall in the region and the entire implementation structure, the construction of the project resulted in a rainwater harvesting system with a self-cleaning disposal system and water treatment using an activated carbon and colloidal silver filter (Figure 17).It is important to emphasise that the aim of installing a rainwater harvesting system is to encompass social, economic and environmental aspects, since during the rainy season the water captured by the roof, which would otherwise be wasted, can be stored and used to cope with the dry months in the region.

Figure 17 - Rainwater harvesting system at a residence in the village of Cuiarana, Salinópolis-PA.
Source: Author (2015).

After implementing the rainwater harvesting project in the village of Cuiarana, Salinópolis-PA, it was necessary to carry out a physical-chemical and bacteriological analysis of the water that was being collected and stored in the reservoir and the water treated by the filter. The water quality assessment before and after passing through the filtration and treatment system was carried out in May 2015 by collecting and analysing the water, the results of which are shown in Table 5.

MAY/2015		
PARAMETERS	CASH	FILTER
Hydrogen Potential (pH)	7,04	6,46
Water Temperature	28 °C	27,7 °C
Turbidity	1.1 µT	0.7 µT
Salinity	0,0 %	0,0 %
Electrical Conductivity	11.8µS	45.7µS
Ammonia (NH3+)	0.19 mg/L	0.42 mg/L
Nitrate (NO3)⁻	1.2 mg/L	1.2 mg/L
Total coliforms	countless	5 CFU/100 ml
Escherichia coli (E. *coli*)	Absent	Absent

Source: Author (2015).

The current standard for the potability of water for human consumption is MS Ordinance No. 2,914 of December 2011, which revoked MS Ordinance No. 518/2004 and provides for the potability standard and procedures for controlling and monitoring the quality of water for human consumption, and CONAMA Resolution No. 357 of 17 March 2005, amended by Resolution 410/2009 and 430/2011, provides for the classification of bodies of water and environmental guidelines for their classification, as well as establishing the conditions and standards for discharging effluents, and other measures. Thus, based on these regulations and studies carried out in this area, they served as the basis for our research.

5. 1Hydrogen Potential   (pH)

It is one of the most important and frequently used parameters in water analysis and must be monitored in order to improve treatment processes and preserve pipes from corrosion or clogging.

The study found that the pH in the water tank was 7.04 and after treatment it was 6.46, showing a slight drop. The current norm regarding water potability standards, Ordinance MS No. 2.914/11, in its Article 39, Paragraph 1ª , proposes that the pH of the water be kept in the range of 6.0 to 9.5, so the parameter analysed complies with the Ministry of Health's potability standards.

In the work by Annecchini (2005), the pH found in the analyses of the rainwater collected by the system ranged from 4.5 to 7.6, showing values that are not in the recommended range. According to the Ministry of Health (2012), water can have a pH outside the recommended range, which does not mean that it is unfit for human consumption.

5. 2Water temperature

Water temperature is the result of solar radiation falling on the water. It has a major influence on biological activities and the growth of organisms; it also determines the types of organisms that inhabit the site, since they have a favoured temperature range in which to develop. If this limit is exceeded, either upwards or downwards, organisms are impacted and more sensitive species can even become extinct from the site.

In the analysis, the temperature data showed values of 28°C **in the water tank and** 27°C after treatment, showing little variation. MS Ordinance No. 2,914/11 lists the temperature and **minimum contact time (in minutes)** with the disinfection process **to be observed during treatment to guarantee the concentration of combined residual chlorine (chloramine) at the treatment outlet, but the study did not carry out this disinfection process.**

5. 3Turbidity

The collection results showed that the turbidity of the water in the water tank was 1.1 μT, while after treatment the water was 0.7 μT. According to MS Ordinance No. 2,914/11, the turbidity standard for post-filtration water is 1.0 μT for treatment through slow filtration, which is the case in this study. As a result, we observed the efficiency of the water treatment in relation to this parameter, where after treatment the turbidity of the water was within the drinking water standards.

Turbidity in water is caused by suspended materials such as clay, silt, finely divided organic and inorganic matter, coloured soluble organic compounds, plankton and other microscopic organisms. The size of suspended particles varies depending on the degree of turbulence in the environment. The presence of these particles causes light to be scattered and absorbed, leaving the water looking cloudy. The transparency **of a natural body of water is one of the main determinants of its condition and** productivity.

In studies carried out in Austin, Texas, Mendez et. al (2011) obtained results similar to those found in this study. The authors assessed the effect of the material of the catchment area on the quality of the rainwater collected, and analysed the turbidity values,

obtained a maximum value of 23 UNT after the first rainwater was diverted to the area covered with tiles.

5.4 Salinity

Salinity is the measure of the quantity of salts in bodies of water. The simplest concept of salinity is the ratio between the total quantity of dissolved ions and the mass of water that serves as their solvent. This parameter is of great importance for characterising bodies of water, as it determines various physical and chemical properties, including the density of the water, the type of fauna and flora and the potential uses of the water.

According to the analysis, this parameter showed the same value both in the reservoir and after treatment, where it was 0.0 per cent. CONAMA Resolution 357/2005 defines water with salinity equal to or less than 0.5% as fresh water, so we characterise the water in this study as fresh water.

5.5 Electrical Conductivity

The analyses showed that the electrical conductivity in the reservoir was 11.8 |iS, and after treatment 45.7 |iS. The increase in conductivity may have been due to the characteristics of the filter present in the treatment, as it is made up of silver colloids, which have the purpose of disinfecting the water, and end up releasing ions, thus causing an increase in this conductivity. According to Companhia Ambiental do Estado de São Paulo - CETESB (2009), electrical conductivity values above 100 |iS indicate impacted environments, so we can see that the water analysed is within the established standards.Electrical conductivity refers to the ability of an aqueous solution to conduct an electric current. Considering that this property depends on the concentration of ions, the higher the ionic concentration, the greater the solution's ability to conduct an electric current, and vice versa. Its practical application is the indication of the degree of mineralisation of water and rapid indication of variations in the concentrations of dissolved minerals.

5. 6Nitrogen series : Nitrate (NO_3^-) and Ammonia (NH_3)$^+$

In the study, the concentration of nitrate (NO_3^-) both in the reservoir and after treatment was 1.2 mg/L, thus causing no variation. Annex VII of Ordinance MS No. 2.914/11 presents the table of drinking water standards for chemical substances that pose a risk to health, including nitrate, whose maximum permitted value is 10 mg/L. Ammoniacal nitrogen corresponds to nitrogen from a compound derived from ammonia. Ammonia is the smallest form of organic nitrogen in water; although it is a small component in the total nitrogen cycle, it contributes to the fertilisation of water as nitrogen is an essential nutrient for aquatic flora and the phytoplankton community. The analysis showed that the concentration of ammonia (NH_3^+) in the reservoir was 0.19 mg/L, while after treatment it was 0.42 mg/L, a significant increase, given that MS Ordinance No. 2,914/11, in its Annex X, in its table of organoleptic drinking standards, sets the maximum permitted value for ammonia at 1.5 mg/L, but the value obtained after treatment is in line with the Ministry of Health's drinking standards.

5. 7Total coliforms

In the analyses, it was possible to observe in the reservoir that the number of heterotrophic bacteria was uncountable on the plate, due to the large quantity present, after going through the treatment process, 5 CFU/100 ml were found, thus occurring a high reduction of bacteria, but the treatment did not manage to completely extinguish the number of bacteria.

Annex I of MS Ordinance 2.914/11 contains a table of microbiological standards for water for human consumption, where total coliforms after treatment must be absent in 100 mL. Article 28 states that the purpose of monitoring heterotrophic bacteria is to verify the integrity of a distribution system, and that the count of heterotrophic bacteria serves, in some way, as a quality control for coliform results. The parameter also fulfils an auxiliary role as an indicator of the stability of the distribution system, and sudden or higher than usual rises should be interpreted as a suspicion that anomalies have occurred.

Counting heterotrophic bacteria provides information on the bacteriological quality of water in a generic way, so heterotrophic bacteria lend themselves to the role of auxiliary indicator of water quality by providing additional information on

possible failures in disinfection, colonisation and biofilm formation in the distribution system, possible changes in water quality or possible problems with the integrity of the distribution system.

Neu (2016)[1] found the presence of total coliforms in rainwater harvesting systems in homes on Ilha das Onças, a riverside region in the state of Pará, where residents did not comply with the recommended collection, cleaning and disposal protocol for the system.

According to the Ministry of Health (2012), the presence of bacteria from the "total coliforms" group in water after treatment is an indication of the inefficiency of the disinfection process used or of (re)contamination problems in the distribution network. On the other hand, this group includes bacteria of non-faecal origin, so this parameter is not considered a good sanitary indicator of water intended for human consumption, nor of raw water, i.e. the presence of these bacteria in raw water does not indicate that it is unfit for consumption.

This makes it necessary to implement measures in the water collection system to reduce or eliminate the presence of these bacteria in drinking water. Firstly, it is necessary to detect the source of the problem so that corrective action can be taken, which can range from changing the pipes in the distribution network to installing

---

[1] Neu, V. (Federal Rural University of Amazonia, UFRA - Campus Belém. Personal communication, 2016)

disinfectant dispensers along the network.

5.8 Escherichia coli (E. coli)

When analysing this parameter, it was possible to see that both in the reservoir and after treatment, the presence of faecal coliforms, in this case E.coli, was absent in 100 mL of the water analysed. According to Article 27 of MS Ordinance No. 2,914/11, drinking water must comply with the microbiological standard set out in Annex I (microbiological standard of water for human consumption). This Annex states that all water intended for human consumption must be free of Escherichia coli. This shows that this parameter is within the established drinking water standards.

CHAPTER 6

FINAL CONSIDERATIONS

The aim of this work was to analyse the use of rainwater in the residence of the Cuiarana Experimental Site, by implementing a rainwater harvesting system, in order to provide an alternative for the community with regard to the problems of drinking water supply, through the collection and treatment of this water.

It was possible to observe that, as this is a social technology, concern about the cost of the project and the feasibility of its implementation was something of great importance in its development. As it is a catchment system that only needs hydraulic devices for its operation, without the use of hydraulic pumps that are expensive to buy and maintain, and still need electricity for their operation, the project sought simple and feasible techniques for the various environments and possible scenarios that it could be inserted in, thus having a low cost compared to other catchment systems.

Most of the parameters showed values in line with the drinking water standards established in MS Ordinance No. 2.914/11. The parameter values of the samples collected from the catchment system at the Sítio residence showed that only total coliforms were outside the standards after water treatment. It was noted that the water treatment was effective in terms of turbidity, since after treatment this parameter was in line with the standards.

Parameters such as electrical conductivity and ammonia ($NH3^{+}$) increased after treatment of the water collected by the system, but this increase did not affect the quality of these parameters, thus keeping them in line with drinking water standards. On the other hand, pH and water temperature showed a decrease in their values, but did not affect the quality of the water. The only parameter that did not change after treatment was nitrate ($NO3^{-}$), but the value also complies with the Ministry of Health regulations.

It was possible to observe that the number of total coliforms in the water in the reservoir was uncountable, but after treatment this number was sharply reduced, with 5 CFU/100 mL. However, Annex I of MS Ordinance 2.914/11 states that total coliforms after treatment must be absent in 100 **mL. The presence of bacteria from the**

"total coliform" group in water after treatment is an indication of the inefficiency of the disinfection used or of contamination problems in the distribution network. It is generally linked to the sanitisation of the system, as the decomposition of organic matter in leaves and branches allows these bacteria to find the nutrients they need to reproduce. As they are not pathogenic organisms, the small presence of these bacteria poses little risk to consumption. Even so, it is necessary to implement corrective methods such as chlorination to reduce or eliminate the presence of these bacteria in the system.

It is important to emphasise that the analysis of the reservoir showed that turbidity was not in line with MS Ordinance No. 2.914/11, but after the water was treated through the filter, this parameter was in line with the drinking water standard, thus showing the effectiveness of the water treatment in relation to turbidity.

One of the main parameters analysed for water potability is the presence of faecal coliforms, especially E. *coli. According* to Article 27 of MS Ordinance No. 2,914/11, as set out in Annex I, any water intended for human consumption must be free of *Escherichia coli* in 100 mL samples. As we have seen in this study, both the water in the reservoir and after treatment is free of faecal coliforms, thus demonstrating the quality of the rainwater.

It should be pointed out that, as this is a pilot project with a single collection, it is very important to monitor the quality of this water over a longer period of time in order to gain a better understanding of the behaviour of the various physical, chemical and bacteriological parameters present in the water collected and treated by the system.

Thus, the search for and implementation of sustainable practices in the Amazon region and the expansion to other regions, through social technologies such as the one presented in this study, aim to improve the quality of life for communities that lack good quality water.

CHAPTER 7

REFERENCES

AMAZONAS. Secretariat for the Environment and Sustainable Development - SDS. Educating to Preserve with Citizenship. In: **Amazon Environmental Exhibition**, 1, 2007. Available at: http://www.sds.am.gov.br/index.php. Accessed on: Dec. 2015.

AMERICAN PUBLISHERS HEALTH ASSOCIATION - APHA. **Standard Methods for The Examination of Water and Wastewater**. 21st ed. Washington DC: American Publishers Health Association, 2005.

ANA - National Water Agency (Brazil). **Atlas Brazil:** urban water supply: national panorama. ANA: Engecorps/Cobrape, Brasília, v.2, 2010. Available at: < http://arquivos.ana.gov.br/institucional/sge/CEDOC/Catalogo/2011/AtlasBrasil-AbastecimentoUrbanodeAgua-PanoramaNacionalv1.pdf>. Accessed on: Dec. 2015.

______. **Training for Singreh**: Monitoring the Water Quality of Rivers and Reservoirs. Brasilia, 2015.

ANNECCHINI, K. P. V. **Harnessing rainwater for non-potable purposes in the city of Vitória (ES).** Dissertation (Master's in Environmental Engineering) - Federal University of Espírito Santo, 2005.

BRAZIL. **National Water Resources Policy.** Law 9433 of 8 January 1997. Establishes the National Water Resources Policy, creates the National Water Resources Management System, regulates item XIX of art. 21 of the Federal Constitution, and amends art. 1 of Law no. 8.001, of 13 March 1990, which amended Law no. 7.990, of 28 December 1989. Available at: < http://www.planalto.gov.br/ccivil_03/Leis/L9433.htm> Accessed on: Dec. 2015.

BRASILIA. **Rainwater Harvesting Programme.** District Law No. 4181/08. Creates the Rainwater Harvesting Programme and makes other provisions. Available at: < http://www.lexml.gov.br/urn/urn:lex:br;distrito.federal:distrital:lei:2008-07-21;4181>. Accessed on: Dec. 2015.

Cáritas Metropolitana De Belém - CAMEBE. **Water at home, clean and healthy.** Belém, PA. Cáritas Metropolitana De Belém, 2007. CD-ROM.

CAMPOS, M. M.; AZEVEDO, F. R. Harnessing Rainwater for Direct Human Consumption. **Electronic journal - Faculdade Integrada Vianna Júnior.** Year V, ed. I, 2013.

CASCAVEL. **Municipal Programme for the Conservation and Rational Use of Water and Reuse in Buildings.** Law No. 4.631/07. Established the Municipal Programme for the Conservation and Rational Use of Water and Reuse in Buildings. Available at: < http://www.acquacon.com.br/9sbcmac/
files/minicurso_2_projeto_de_aproveitamento_e_c

aptacao_de_agua_de_chuva_no_meio_urbano.pdf >. Accessed on: Dec. 2015.

COMPANHIA AMBIENTAL DO ESTADO DE SÃO PAULO. **Guia Nacional de Coleta e Preservação de Amostras: Água, Sedimento, Comunidades Aquáticas e Efluentes Líquidos.** BRANDÃO, C. J.; BOTELHO, M. J. C.; SATO, M. I.; LAMPARELLI, M. C. São Paulo. CETESB: Brasília. ANA. 2011. 326p.

______. **Inland Water Quality in the State of São Paulo:** Environmental and Sanitary Significance of Water and Sediment Quality Variables and Analytical and Sampling Methodologies. São Paulo State Government, São Paulo, 2009.

CURITIBA. **Programme for the Conservation and Rational Use of Water in Buildings - PURAE.** Law No. 10785, of 18 September 2003. Creates, in the municipality of Curitiba, the Programme for the Conservation and Rational Use of Water in Buildings - PURAE. Available at: < http://www.jusbrasil.com.br/topicos/15693346/lei-n-10785-de-18-de-setembro-de- 2003-do-municipio-de-curitiba>. Accessed on: Dec. 2015.

FERNANDES, A. L. G. **"SUSTAINABILITY OF CONSTRUCTIONS" Building for a better future - Reusing water.** Monograph (Specialisation in Civil Construction) - Federal University of Minas Gerais, Belo Horizonte, 2009.

FIGUEIREDO, L. H. S. **Utilisation of Rainwater for Non-Potable Purposes, Analysis in a Single Family Residence in Macaúbas-Ba.** Final course work (Civil Engineering) - Catholic University of Salvador, Salvador, 2014.

FORTLEV. **Polythene water tanks.** Available at   :< http://www.fortlev.com.br/produto/caixa-dagua-de-polietileno-2/>. Accessed on: Dec. 2015.

GOMES, M. A. F. Água: **sem ela seremos o planeta Marte de amanhã.** [S. l.] Embrapa, 2011. Available at: <http://www.cnpma.embrapa.br/down_hp/464.pdf>. Accessed on: Dec 2015.

GONÇALVES, C. C. **Harnessing rainwater for rural water supply in the Amazon. Case study: Grande and Murutucú islands, Belém-PA.** Dissertation (Master's in Civil Engineering), Federal University of Pará, Belém, 2012.

GUARULHOS. **Municipal Programme for the Rational Use of Water.** Municipal Law No. 6.511/09. Institutes Rainwater Utilisation. Available at: < http://www.jusbrasil.com.br/diarios/98865552/dom-gru-legal-28-08-2015-pg-28?ref=topic_feed>. Accessed on: Dec. 2015.

HAGEMANN, S. E. **Evaluation of Rainwater Quality and the Feasibility of its Collection and Use.** Master's dissertation (Postgraduate Programme in Civil Engineering, Area of Concentration in Water Resources and Environmental Sanitation). Federal University of Santa Maria (UFSM, RS), p.141, 2009.

HESPANHOL, I. Potential for water reuse in Brazil: agriculture, industry, municipalities, aquifer recharge. **BAHIA ANÁLISE & DADOS** Salvador, v. 13, n. ESPECIAL, p. 411437, 2003.

BRAZILIAN INSTITUTE OF THE ENVIRONMENT AND RENEWABLE NATURAL RESOURCES - IBAMA. **History2008.** Available at< http://www.ibama.gov.br/patrimonio/>. Accessed on: Dec. 2015.

BRAZILIAN INSTITUTE OF GEOGRAPHY AND STATISTICS - IBGE. **National Basic Sanitation Survey 2008.** Available at < http://www.ibge.gov.br>. Accessed on: Dec. 2015.

LIMA, R. P.; MACHADO, T. G. **Harnessing Rainwater: Analysing the Cost of Implementing the System in Buildings.** Course Conclusion Paper (Civil Engineering) - Barretos Educational Foundation University Centre, Barretos, 2008.

LIMA,R.T.; SOUZA, Paulo Jorge de Oliveira. Ponte de. **Radiometric Relationships in a Mango Orchard, cv. Tommy Atkins, in the Northeast of Pará.** Dissertation (Master's Degree), Federal Rural University of Amazonia, Belém, 2012.

MATOS, B. A. et al. Availability and demands of water resources in the 12 hydrographic regions of Brazil. 17th Water Resources Symposium, São Paulo, 2007. **Proceedings of the 17th Water Resources Symposium.** São Paulo, 2007.

MENDEZ, C. B.; KLENZENDORF, J. B.; AFSHAR, B. R.; SIMMONS, M. T.; BARRETT, M. E.; KINNEY, K. A.; KIRISITS, M. J.. **The effect of roofing material on the quality of harvested Rainwater.** Water Research, v. 45, n. 5, p. 2049 - 2059, 2011.

MINISTRY OF HEALTH. Norms and Standards for the Potability of Water Intended for Human Consumption. **ORDINANCE N° 2.914 of 12/12/2011,** Brazil. Provides for procedures to control and monitor the quality of water for human consumption and its potability standard.

______. **Frequently asked questions about MS Ordinance No. 2.914/2011.** Brasilia, 2012.

OLIVEIRA, D. R. C. **Harnessing Rainwater in the Amazon.** Belém, Federal University of Pará: Postgraduate Programme in Civil Engineering, 2009. Project: MCT/CNPq/CT-HIDRO n. 21.

UNITED NATIONS ORGANISATION. UN releases 4th **World Water Forum report on water resources.** VEJA. April: 2012. Available at: <http://veja.abril.com.br/noticia/ciencia/onu-apresenta-relatorio-sobre-recursos-hidricos-em- forum-mundial-da-agua>. Accessed on: Dec. 2015.

PETERS, M.R. **Potential use of alternative water sources for non-potable purposes in a residential unit.** Dissertation (Master's in Environmental Engineering) - Federal University of Santa Catarina, Florianópolis, 2006.

PORTO ALEGRE. **Programme for the Conservation, Rational Use and Reuse of Water.** Law No. 10.506/2008. Establishes the Water Conservation, Rational Use and Reuse Programme . Available at :< http://www.mprs.mp.br/atuacaomp/not_artigos/id17232.html>. Accessed on: Dec. 2015.

REBOUÇAS, A. C.; BRAGA, B.; TUNDISI, J. G. **Águas doces no Brasil:** Capital ecológico, uso e conservação. Escrituras ed., São Paulo 1999.

RIO DE JANEIRO. Decree no. 23.940, of 30 January 2004. Makes it compulsory, in the cases provided for, to adopt reservoirs that allow rainwater runoff to be delayed in the drainage network. Available at: < http://www0.rio.rj.gov.br/smac/up_arq/ DEC-23940-04-aguaspluv.pdf>. Accessed on: Dec. 2015.

RODRIGUES, J. C. **Energy balance and phenological behaviour in mango orchards in the Northeast of Pará.** Dissertation (Master's in Agronomy) - Federal Rural University of Amazonia, 2012.

ROSA, R. G. **Utilisation of rainwater for drinking - case study: municipality of Belém-PA.** 2011. Dissertation (Master's in Civil Engineering), Federal University of Pará, Belém, 2011.

SÃO PAULO, Law No. 12.526, of 2 January 2007. Establishes standards for flood containment and rainwater disposal. Available at: < http://www.al.sp.gov.br/ repositorio/legislação/lei/2007/lei-12526-02.01.2007.html>. Accessed on: Dec. 2015.

SOUZA, R. O. R M.; SCARAMUSSA, P. H. M.; AMARAL, M. A. C. M.; PEREIRA NETO, J. A.; PANTOJA, A. V.; SADECK, L. W. R.. Heavy rainfall equations for the state of Pará. **Journal of Agricultural and Environmental Engineering**, v. 16, n. 9, p. 999-1005, 2012.

STÉFANI. **Sterilising candle.** Available at: <http://www.ceramicastefani.com.br/acessorio- stefani/acessorio/3/vela-sterilizante. Accessed on: Dec. 2015.

TASSI, R. **Runoff Control at Source:** Rainwater Harvesting and Utilisation Project in the Urban Environment. Minicourse, Federal University of Santa Maria - RS, Santa Maria, 2014.

TESTON, A. **Rainwater Harnessing: A Qualitative Study of the Main Systems.** Monograph (Specialisation in Sustainable Construction). Federal Technological University of Paraná - UTFPR. Curitiba, 2012.

TOMAZ, P. **Harnessing Rainwater.** ed. 2. São Paulo: Navegar. 2003

______. **Harnessing rainwater in urban areas for non-potable purposes.** Pontifical Catholic University of Campinas. Campinas, 2009.

______. **Harnessing rainwater for urban areas and non-potable purposes**. São Paulo: Navegar Editora, p. 180, 2003.

TORDO, O. C. **Characterisation and evaluation of the use of rainwater for drinking purposes.** Dissertation (Master's in Environmental Engineering) - Centre for Technological Sciences and Postgraduate Programme in Environmental Engineering, Regional University of Blumenau. Blumenau, 2004.

UFPE. **DESVIUFPE.** Available at: <
https://www.ufpe.br/lea/index.php?option=com_content&view=article&id=309:baixe
-aqui- the-videos-about-deviufpe&catid=2:course&Itemid=122>. Accessed on: Dec. 2015.

UFRA News Today. **Ufra project is a finalist in the 2014 Ana Prize**. Issue no. 198, 26 November 2014.

VELOSO, N.S.L. **Rainwater and local development; the case of supplying the islands of Belém**. Dissertation (Master's in Natural Resource Management and Local Development in the Amazon), Environment Centre, Federal University of Pará, Belém, 2012.

______. **Rainwater for Supply in the Amazon.** Revista Movendo Ideias. v. 17, n. 1, 2012.

VON SPERLING, M. **Princípios do Tratamento Biológico de Águas Residuárias:** Introdução à qualidade das águas e ao tratamento de esgotos. v. 1. 3. ed. Department of Sanitary and Environmental Engineering (DESA). Federal University of Minas Gerais (UFMG). 2005.

SHIKLOMANOV, I. A. International Hydrological Programme - IHP - IV/UNESCO, 1998. In: Rebouças, A. C. et al. **Águas doces no Brasil:** Capital ecológico, uso e conservação. Escrituras, 1999.

I want morebooks!

Buy your books fast and straightforward online - at one of world's fastest growing online book stores! Environmentally sound due to Print-on-Demand technologies.

Buy your books online at
www.morebooks.shop

Kaufen Sie Ihre Bücher schnell und unkompliziert online – auf einer der am schnellsten wachsenden Buchhandelsplattformen weltweit! Dank Print-On-Demand umwelt- und ressourcenschonend produziert.

Bücher schneller online kaufen
www.morebooks.shop

Printed by Books on Demand GmbH, Norderstedt / Germany